Introduction

Solar eclipses have fascinated humanity for millennia, captivating our imagination and inspiring awe.

These celestial events occur when the moon passes between the sun and the Earth, temporarily casting a shadow on our planet.

While the solar eclipse is a breathtaking spectacle to behold, its impact extends far beyond mere visual splendor.

In this comprehensive exploration, we delve into the profound effects of solar eclipses on Earth and its diverse ecosystems, from the physical to the behavioral adaptations of various organisms.

Understanding Solar Eclipses

Before delving into its effects, it's essential to understand the mechanics of a solar eclipse.

A solar eclipse occurs when the moon's orbit intersects with the plane of Earth's orbit around the sun, resulting in the alignment of the sun, moon, and Earth in a specific configuration.

There are three types of solar eclipses: total, partial, and annular, each characterized by the extent to which the sun is obscured by the moon.

During a total solar eclipse, the moon completely covers the sun, plunging the surrounding areas into darkness for a brief period.

A partial solar eclipse occurs when only a portion of the sun is obscured by the moon, while an annular eclipse happens when the moon covers the center of the sun, leaving a ring of sunlight visible around the moon's silhouette.

Physical Effects of Solar Eclipses

Solar eclipses have both direct and indirect physical effects on Earth's environment.

The sudden reduction in solar radiation during a total solar eclipse leads to a noticeable drop in temperature in the affected regions.

This phenomenon, known as eclipse cooling, can cause a decrease of several degrees Celsius within minutes, altering local weather patterns temporarily.

Furthermore, the abrupt decrease in sunlight during a solar eclipse affects photosynthesis in plants.

As plants rely on sunlight for energy production, the reduction in solar radiation can temporarily slow down photosynthetic activity, impacting plant growth and metabolism.

However, studies have shown that most plants can recover quickly once sunlight returns to normal levels.

Solar eclipses also influence atmospheric conditions, leading to changes in wind patterns and atmospheric pressure.

These alterations can have cascading effects on weather systems, potentially affecting cloud formation, precipitation, and atmospheric circulation on both local and regional scales.

Ecological Impacts of Solar Eclipses

The ecological impacts of solar eclipses extend beyond mere changes in temperature and weather patterns.

These celestial events can disrupt the behavior and physiology of various organisms, prompting unique responses and adaptations.

One notable example is the behavior of diurnal animals during a solar eclipse.

Diurnal species, such as birds, insects, and mammals, are accustomed to the daily rhythms of sunlight and darkness.

The sudden onset of darkness during a total solar eclipse can trigger instinctual responses in these animals, including changes in foraging behavior, vocalizations, and roosting activities.

Similarly, nocturnal animals may exhibit altered behavior during a solar eclipse, despite being active during darkness.

The sudden disruption of natural light cues can confuse nocturnal species, leading to changes in their activity patterns and interactions with their environment.

Marine ecosystems are also influenced by solar eclipses, particularly in coastal regions where the eclipse's effects are most pronounced.

The temporary decrease in sunlight can affect the behavior of marine organisms, including plankton, fish, and marine mammals.

For example, some species of plankton may alter their vertical migration patterns in response to changes in light intensity, which can have ripple effects throughout the marine food web.

Additionally, terrestrial and aquatic plants may experience changes in their growth and reproductive cycles due to altered light conditions during eclipses. For instance, some species of flowering plants may delay or accelerate blooming in response to the sudden decrease and increase in sunlight.

Cultural and Societal Significance

Beyond its scientific implications, the solar eclipse holds cultural and societal significance for many civilizations around the world.

Throughout history, solar eclipses have been interpreted as omens, symbols of divine intervention, or celestial phenomena imbued with mystical significance.

Ancient civilizations often viewed solar eclipses as harbingers of change or symbols of cosmic conflict between celestial forces.

In some cultures, solar eclipses were believed to portend disasters, wars, or the wrath of vengeful deities.

Conversely, other societies celebrated solar eclipses as auspicious events, marking the alignment of cosmic energies or the renewal of the natural world.

Today, solar eclipses continue to capture the public's imagination, drawing millions of spectators to witness these rare celestial events firsthand.

Eclipse chasers travel the globe in pursuit of the perfect viewing location, while scientists seize the opportunity to conduct research and study the sun's corona during total solar eclipses.

Solar eclipses are celestial phenomena that transcend mere astronomical events, exerting profound effects on Earth and its diverse ecosystems.

From the physical changes in temperature and weather patterns to the behavioral adaptations of organisms, solar eclipses leave a lasting imprint on the natural world.

Moreover, the cultural and societal significance of solar eclipses underscores humanity's enduring fascination with these cosmic spectacles.

As we continue to study and marvel at the mysteries of the universe, solar eclipses serve as reminders of our place within the vastness of space and time, connecting us to the rhythms of the cosmos in ways both profound and sublime.

The 2024 Solar Eclipse: Anticipation and Impact

As the celestial clock ticks forward, humanity eagerly awaits the next grand spectacle in the heavens – the solar eclipse of 2024. Scheduled to grace the skies on April 8, 2024, this event promises to captivate observers across North America and beyond. With its path of totality stretching from Mexico to Canada, the 2024 solar eclipse holds significant scientific, cultural, and societal importance.

Anticipation and Preparation

Excitement has been building among astronomers, enthusiasts, and the general public alike as the date of the 2024 solar eclipse draws near.

Unlike some celestial events that may be visible only from remote locations, this eclipse offers accessibility to millions across North America. Cities such as Dallas, Indianapolis, Cleveland, and Buffalo lie within the path of totality, providing ample opportunities for viewing and study.

Preparation for this rare event is already underway, with astronomers and researchers gearing up to capture valuable data.

Advanced imaging technologies, including high-resolution telescopes and sophisticated cameras, will be deployed to observe the intricate details of the sun's corona during totality.

Citizen scientists and amateur astronomers are also eagerly planning their observations, equipped with solar viewing glasses and telescopes to witness the spectacle firsthand.

The 2024 solar eclipse presents a golden opportunity for scientists to deepen their understanding of the sun and its surrounding environment.

During totality, when the moon completely obscures the sun, researchers will have a rare chance to study the sun's outer atmosphere, known as the corona.

This region, normally invisible against the sun's brilliant glare, holds key insights into solar activity, including the mechanisms driving solar flares and coronal mass ejections.

Moreover, the 2024 eclipse provides a unique natural laboratory for studying Earth's ionosphere and atmosphere.

Changes in temperature, pressure, and electromagnetic conditions during the eclipse offer valuable data for atmospheric scientists, aiding in the refinement of climate models and weather prediction algorithms.

Cultural and Societal Impact

Beyond its scientific significance, the 2024 solar eclipse carries profound cultural and societal implications.

Across history, solar eclipses have inspired awe and wonder, fueling myths, legends, and religious interpretations. In modern times, they continue to serve as opportunities for communal gatherings and shared experiences.

For many communities along the eclipse's path, the event holds economic significance as well.

Hotels, restaurants, and tourism operators are gearing up to accommodate the influx of visitors eager to witness the spectacle.

Local businesses may see a surge in revenue as eclipse watchers flock to their cities, bringing with them enthusiasm and excitement.

Safety Precautions

While the solar eclipse offers a dazzling display of cosmic beauty, it's essential to observe it safely.

Staring directly at the sun, even during an eclipse, can cause permanent eye damage or blindness.

Specialized solar viewing glasses or handheld solar filters must be used to protect the eyes while observing the eclipse.

Furthermore, observers should exercise caution when traveling to view the eclipse, ensuring they have adequate supplies, transportation, and accommodations.

Planning ahead and adhering to local regulations and safety guidelines can help ensure a memorable and safe viewing experience for all.

Conclusion

The 2024 solar eclipse promises to be a momentous event, uniting communities and sparking scientific curiosity across North America.

As the moon's shadow sweeps across the continent, observers will gather to witness nature's grand spectacle, pondering the mysteries of the cosmos and their place within it.

From scientific research to cultural celebrations, the 2024 solar eclipse serves as a reminder of the profound connections between Earth and the cosmos, inspiring wonder and awe in all who behold it.

Thank you for reading...

www.ingramcontent.com/pod-product-compliance
Lightning Source LLC
Chambersburg PA
CBHW030108230526
45471CB00003B/1310